Genre > **Nonfiction**

Essential Question
What information have scientists learned from the *Galileo* spacecraft?

The *Galileo* Mission to Jupiter

by Peter Rader

Chapter 1
The Journey Begins 2

Chapter 2
The Real Galileo 4

Chapter 3
Along the Way 6

Chapter 4
Exploring Jupiter 8

Chapter 5
More Discoveries 10

Respond to Reading 12

PAIRED READ Probe 13

Glossary 15

Index 16

Chapter 1
The Journey Begins

When the Space Shuttle *Atlantis* leapt into the sky on October 18, 1989, it began a journey of nearly 3 billion miles. The *Galileo* (GAL-uh-lee-oh) spacecraft which *Atlantic* carried would make the trek to Jupiter. Hours after liftoff, the shuttle crew put *Galileo* into space.

Scientists believed that this **mission** was going to be exciting. Jupiter is the biggest planet in our **solar system**. It has more than 60 moons! Some of those moons have large volcanoes. Others have deep oceans. The *Galileo* spacecraft would fly closer to Jupiter than ever before. It would get close enough to take pictures and test the planet's **atmosphere**. The results of the mission were even more fascinating and valuable than scientists expected.

The *Galileo* spacecraft is put into space.

The *Galileo* Spacecraft

Although *Galileo* was an unmanned craft, nearly 1,000 people worked on its mission. During the journey, the National Aeronautics and Space Administration (NASA) sent more than 275,000 commands to the spacecraft. *Galileo* sent about 14,000 photographs to Earth.

Galileo weighed more than 2,200 kilograms (2. tons). It was about 5 meters (17 feet) high. Its boom arm could reach about 11 meters (36 feet). NASA built the craft with tough, light, high-tech materials similar to what you would find in a fighter jet. Beryllium and aluminum formed parts of the structure, and the booms were made of carbon composites. The spacecraft could tolerate temperature extremes of hundreds of degrees below and above freezing without losing function.

Workers at NASA's mission control kept track of Galileo all along its way.

Chapter 2
The Real Galileo

The Galileo spacecraft was named after the Italian astronomer Galileo Galilei (GAL-uh-lee-oh gal-uh-LAY-ee), who was born in Pisa in 1564. The Dutch made telescopes that made an object look three times closer. After Galileo heard about the invention, he designed and built a telescope that brought objects twenty times closer.

Galileo demonstrating his telescope

With his telescope, Galileo was the first to see four of Jupiter's moons. He also observed dark patches on the Sun, called sunspots. He saw the mountains of Earth's Moon, too.

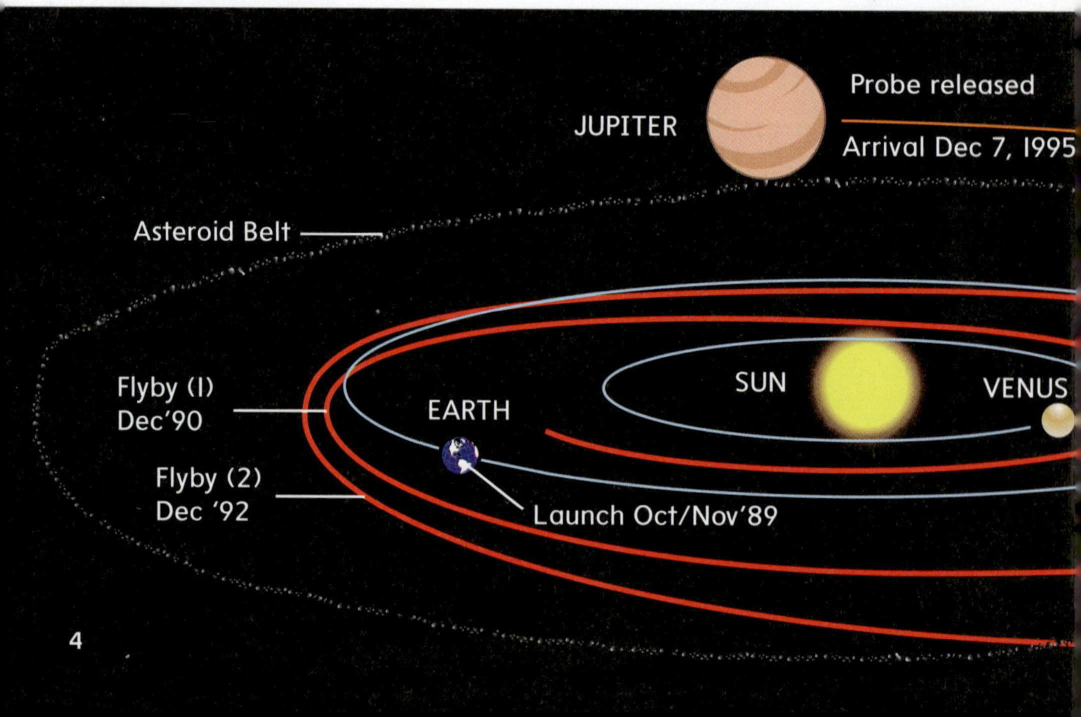

Galileo observed that Venus changed phase like Earth's Moon. The changing shadows proved that Venus went around the Sun. Galileo thought the Earth orbited the Sun, and not the opposite. He was right.

Nearly 400 years later, the *Galileo* spacecraft used the orbit of Venus in an interesting way. Instead of flying directly to Jupiter, the spacecraft used Venus's gravity to "slingshot" back toward Earth before going toward Jupiter.

By using other planets' gravity, *Galileo* built up the speed it would need to reach the outer planets. It would travel great distances through the solar system using very little fuel.

This is the flight path of then Galileo spacecraft, from its launch in 1989 to its arrival at Jupiter in 1995.

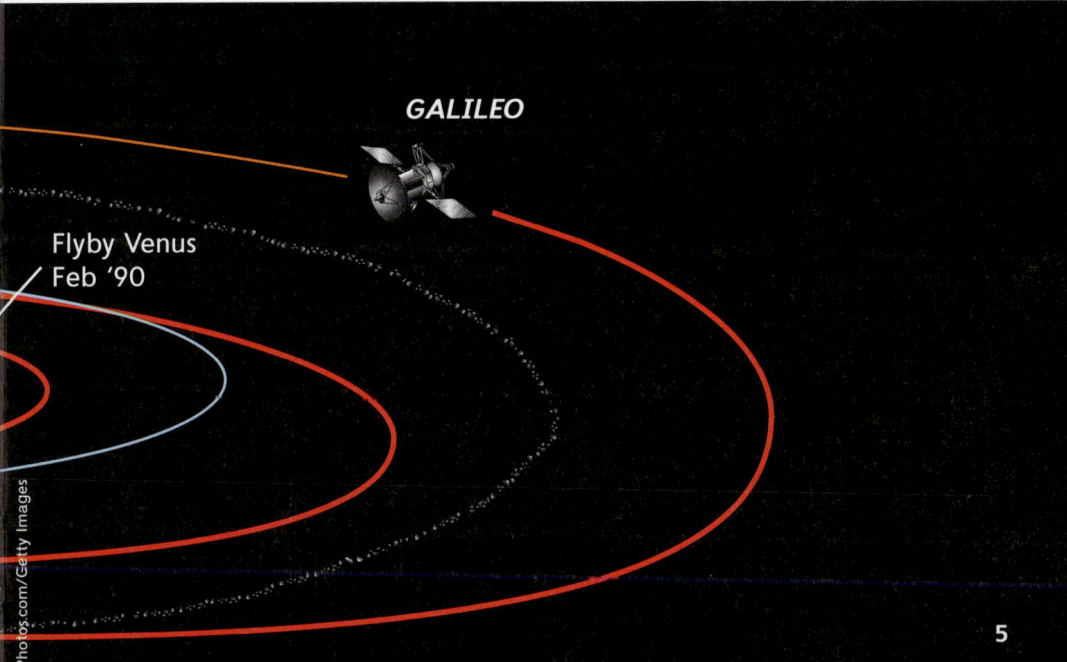

Chapter 3
Along the Way

In its journey of almost 3 billion miles, *Galileo* passed some extraordinary sights and created several "firsts."

It made close **flybys** past the **asteroids** Gaspra and Ida. The spacecraft's cameras discovered that the asteroid Ida had its own moon, which scientists named Dactyl. This was the first time any asteroid had been found to have a moon. The pictures that *Galileo* sent back to Earth were the best asteroid photographs ever taken.

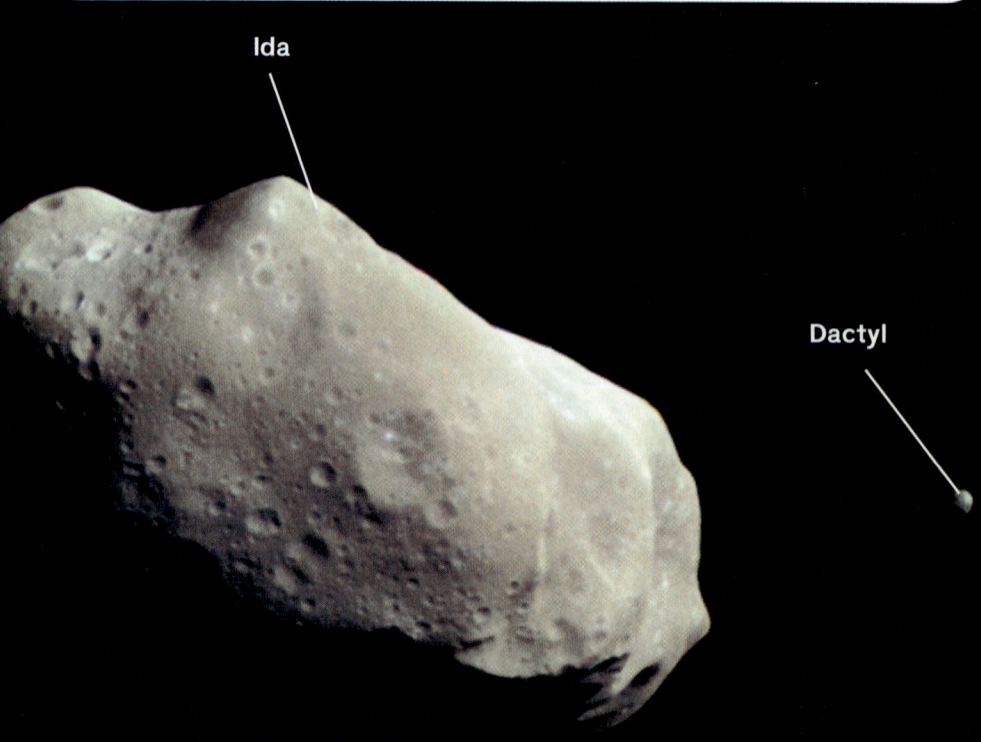

Photographs of the asteroid Ida and its moon Dactyl were taken by *Galileo* on August 28, 1993.

The Hubble Space Telescope took pictures of Jupiter and the comet Shoemaker-Levy 9.

As *Galileo* made its final approach to Jupiter, another object was heading to the planet: the newly discovered comet Shoemaker-Levy 9. The comet collided with Jupiter in July 1994, and *Galileo's* cameras recorded the event. The photos are the only observations human beings have ever made of a comet striking a planet in real time.

Chapter 4
Exploring Jupiter

On December 7, 1995, after more than six years, *Galileo* arrived at Jupiter.

Jupiter is both the largest and the fastest rotating planet in the Solar System. Jupiter's fast spinning causes great winds and produces the bands of clouds that cover the planet. The wild atmosphere also created the famous Great Red Spot, a huge storm more than three hundred years old.

Galileo's mission was to spend two years exploring Jupiter and its many moons. To do so, NASA had to slow down the spacecraft to place *Galileo* into orbit around Jupiter. Never before had a spacecraft gone into orbit around one of the outer planets.

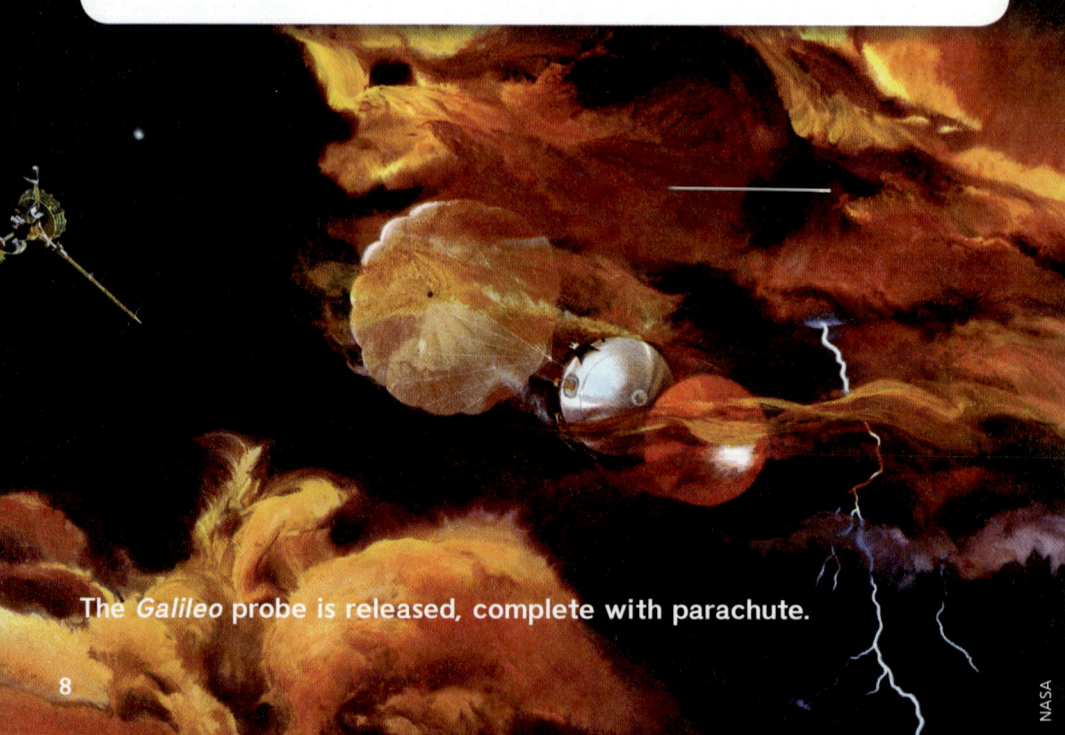

The *Galileo* probe is released, complete with parachute.

The *Galileo* probe is shown arriving near Jupiter.

One of *Galileo's* tasks was to release a special **probe** that would test Jupiter's atmosphere. Galileo dropped the probe over Jupiter with a parachute.

The probe discovered that Jupiter's wind speeds are greater than 644 kilometers (400 miles) an hour. Lightning strikes on Jupiter are ten times stronger than Earth's. Jupiter's atmosphere is one quarter helium, much like the Sun's. If Jupiter were much larger, it would be a star.

During its orbits around Jupiter, *Galileo* came close to some of the giant planet's moons. At last count, Jupiter had at least 61. Four moons—discovered by Galileo Galilei—are the size of our Moon or bigger. Others are smaller.

Chapter 5
More Discoveries

Jupiter's closest moon, Io, has many volcanoes. *Galileo* saw exploding volcanoes more than 320 kilometers (200 miles) high!

Galileo's sensors also discovered that three other moons, Europa, Callisto, and Ganymede, possibly have oceans below the surface. Europa's ocean might even hold more water than exists on all of Earth. Scientists think it is possible these moons might contain early life forms.

Jupiter's largest moon, Ganymede, is the size of a small planet. It is bigger than Mercury!

These pictures show four of Jupiter's moons.

Galileo burnt up in Jupiter's atmosphere.

Galileo's discoveries were awesome! NASA had the craft orbit Jupiter 34 times in eight years. The spacecraft lasted much longer than anyone had expected.

After almost 3 billion miles, *Galileo* had used only 931 liters (246 gallons) of fuel. It would be forced to crash once it ran out of fuel.

Because *Galileo* had discovered there might be life on Jupiter's moons, NASA wanted to avoid infecting the moons. It decided to crash *Galileo* into Jupiter. There it would burn and break up without harming the thick atmosphere.

On September 21, 2003, *Galileo* crashed into Jupiter. It disappeared forever into Jupiter's atmosphere. It was a noble end for the little spacecraft.

Summarize

Use important details from The Galileo Mission to Jupiter to summarize the selection. Your graphic organizer may help you.

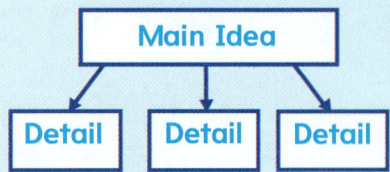

Text Evidence

1. Describe some of the discoveries made by the Galileo spacecraft. How might some of the discoveries be important for humans someday? Explain.

2. Reread Chapter 4 with a group. What is the main idea of Chapter 4? What are key details that support the main idea? MAIN IDEA AND KEY DETAILS

3. What does *changed phase* mean on page 5? Which context clues helped you figure out the meaning? CONTEXT CLUES

4. Galileo Galilei is sometimes called the "Founder of Modern Science." Find out why. Do research about his life and work. Then use what you learned to write his life story. Include key details about some of the scientific discoveries he made. WRITE ABOUT READING

Compare Texts

Read how Jennae learns about guiding a space probe to Jupiter.

Probe

Boring! That was the word that came to mind when Jennae was thinking about her trip to the planetarium.

"What's so great about watching stars?" Jennae asked Vanessa, her best friend. "If I want to look at stars, all I need to do is go outside at night."

Jennae walked into the planetarium with her class. She prepared for the boredom that she knew was coming, when her class unexpectedly entered another room.

A planetarium manager stood in front of the class "Today, you are going to be the first class to take part in a program called Mission to Jupiter."

Jennae's curiosity grew. Perhaps, a planetarium was not only about stargazing.

Jennae partnered with Vanessa to use a computer-based program. It allowed them to simulate navigating a space probe to Jupiter. They launched the probe from an old-fashioned space shuttle. They guided the probe past Mars and then maneuvered through the asteroid belt, nearly avoiding a collision with a huge asteroid.

"This is kind of fun," Jennae sheepishly admitted to her friend at her side.

Once out of the belt, they prepared the probe to explore Jupiter. They rocketed by one of the countless moons of Jupiter, coming close enough to take pictures of it. Then it was onto Jupiter, where they took pictures of the giant planet and scanned it for life. They found no life. So, they placed the probe in orbit around the planet to complete their mission.

"Wow," Jennae said when the program ended. She turned to Vanessa, "I guess planetariums aren't so boring after all."

Make Connections

What convinced Jennae to change her mind about the planetarium? TEXT TO TEXT

Glossary

asteroid *(AS-tuh-royd)* chunks of rocks and metal that circle the Sun *(page 6)*

astronomer *(uh-STRON-uh-muhr)* a scientist who studies the Sun, Moon, stars, and other planets *(page 4)*

atmosphere *(AT-muhs-feer)* gases that surround a planet or moon *(page 8)*

comet *(KOM-it)* a chunk of ice and rock that orbits the Sun in a long, narrow path *(page 7)*

flyby *(FLYE-bye)* flying by a planet without landing or orbiting it *(page 6)*

mission *(MISH-uhn)* a special task for a spacecraft to perform *(page 2)*

orbit *(AWR-bit)* the path an object follows as it moves in a circle around something else *(page 5)*

outer planet *(OW-tuhr PLAN-it)* one of the four planets farther from the Sun—Jupiter, Saturn, Uranus, or Neptune *(page 8)*

probe *(PROHB)* a spacecraft used for exploring a planet *(page 9)*

solar system *(SOH-luhr SIS-tuhm)* the Sun and all the objects that go around it *(page 2)*

Index

asteroid, *6*

Atlantis, *2*

atmosphere, *8–9, 11*

Callisto, *10*

Dactyl, *6*

Earth, *4–6, 9*

Europa, *10*

flight path, *5*

flyby, *6*

Galilei, Galileo, *4–5, 9*

Galileo, *2–11*

Ganymede, *10*

Great Red Spot, *8*

Ida, *6*

Io, *10*

Jupiter, *2, 4–5, 7–11*

Kennedy Space Center, *2*

Mercury, *10*

mission control, *3*

moons, *2, 4, 8–11*

National Aeronautics and Space Administration (NASA), *3, 8, 11*

ocean, *2, 10*

outer planets, *5, 8*

probe, *9*

Shoemaker-Levy 9, *7*

solar system, *2, 5*

Sun, *4–5, 9*

telescope, *4*

Venus, *5*

volcano, *2, 10*